SUICIDAL PILOTS, ARE THEY?

Mi Ko

With so many airplane crashes attributed to suicidal pilots, one wonders whether indeed the crashes were caused by suicidal pilots or something else. Careful analysis of a number of suspiciously-explained crashes leads to a rather unexpected and scary conclusion.

2018/3/17 edition

Readers' feedback including but not limited to typos, mistakes, corrections, suggestions, participation in research, donations to research (these would be greatly appreciated), etc. may be communicated to the author by email to outercorestudies@gmail.com, the author may not be able to reply to all emails but will try to reply to as many as possible.

Contents

1   Airplane crashes in the Alps.

> Once is an accident,   twice is a coincidence,
> three times is a pattern.   Ian Fleming.
>
> The greatest obstacle to discovery is not igno-
> rance, it is the illusion of knowledge. Author
> unknown.

It has been three years since the tragic crash of Germanwings 9525. The cause of the crash was ruled to be co-pilot's suicide. A huge media frenzy[1] followed the announcement with numerous 'experts' in aviation, psychology, and every other imaginable field offering their analysis and advice. What has not been mentioned by the media is that the crash was only one of at least six mysterious airplane crashes in the same geographical region:

1) Germanwings 9525 crashed on 2015/3/24 at 9:41 at $44.28^oN, 6.44^oE$ [2,3]. After an unprecedentedly short investigation of less than two days[4], it was announced that the mentally sick 28-year-old co-pilot Andreas Lubitz barricaded himself inside the cockpit and deliberately rammed the plane into the Alps. But why would the physically fit and smiling 28-year old co-pilot running marathons, as shown all over the Internet, who just purchased a brand new car, commit suicide? Those who

---

[1]References are provided to sources available at the time of writing of this paper. The author attempted to verify the information against other sources whenever possible as both reputable and not-so-reputable sources may provide unreliable information. Even supposedly reliable sources may not be reliable, e.g. Nayirah testimony before the Congressional Human Rights Caucus, `https://en.wikipedia.org/wiki/Nayirah_(testimony)`; Colin Powell's presentation to the United Nations Security Council about the weapons of mass destruction in Iraq, `https://en.wikipedia.org/wiki/United_Nations_Security_Council_and_the_Iraq_War`; Pan Am 103 controversy with one of only two accused, Lamin Khalifah Fhimah, acquitted by the court, and the conviction of the other one mired in uncertainty, `https://www.victimsofpanamflight103.org/`; `http://www.aljazeera.com/indepth/features/2016/02/pan-flight-103-lockerbie-bomber-guilty-160211101436307.html`; `https://en.wikipedia.org/wiki/Pan_Am_Flight_103_conspiracy_theories`; `http://edition.cnn.com/2013/09/26/world/pan-am-flight-103-fast-facts/`; the controversy of Air India 182 with two accused, Malik and Bagri, acquitted and one, Reyat, accused of merely a perjury. Even so-called scientific sources may be quite unreliable, e. g. USGS recently changed its earthquake database `https://earthquake.usgs.gov/earthquakes/search/` essentially admitting that the previous database was significantly flawed.

[2]All coordinates in text are rounded off to two decimal places, coordinates in the pictures may be given with higher precision. All time is in UTC unless otherwise stated.

[3]`http://en.wikipedia.org/wiki/Germanwings_Flight_9525`

[4]`http://www.ibtimes.co.uk/andreas-lubitz-white-christian-germanwings-crash-pilot-should-be-called-terrorist-1493716`; `http://edition.cnn.com/2015/03/26/europe/germanwings-plane-crash-pilots/` are dated 2015/3/26. According to `http://www.ibtimes.com/germanwings-pilot-andreas-lubitz-was-framed-friends-allege-lufthansa-crash-cover-1864914` antidepressants prescribed to treat a serious "psychosomatic illness" were found in the co-pilot's home.

knew him insisted Lubitz was not suicidal[5] nor do many pilots buy the story[6]. Many witnesses reported explosion and smoke before the plane plunged into the Alps and debris was found upstream from the crash site suggesting that at least one piece of fuselage had "been detached from the aircraft before impact"[7]; which can only happen if the plane exploded in the air.

2) Air France 178 crashed on 1953/9/1 at 23:30 Paris time at practically the same place as the previous flight at $44.29^{\circ}N, 6.7^{\circ}E$ [8] amidst violent storms after "the flight had deviated from the planned course for unknown reasons".

3) Crossair 498 crashed on 2000/1/10 at 16:54 at $47.47^{\circ}N, 8.47^{\circ}E$ [9]. The investigation concluded that "... pilot Pavel Gruzin's body revealed traces of the tranquilizer Phenazepam ... an open packet of the Russian-made drug in baggage belonging to Gruzin ... commander took the aircraft into a spiral dive to the right because ... he had lost spatial orientation ... "[10].

4) Swissair 330 exploded on 1970/2/21 at 12:15 at $47.54^{\circ}N, 8.24^{\circ}E$ [11], very close to the previous crash site. The explosion was attributed to a bomb; yet, the only time in history, no one has ever claimed responsibility for the bombing nor have the perpetrators ever been named.

5-6) Air India 245 crashed on 1950/11/3 at 9:43 am at $45.83^{\circ}N, 6.86^{\circ}E$ amidst stormy weather[12] and Air India 101 crashed on 1966/1/24 sometime after 7:00 am at practically the same place at $45.88^{\circ}N, 6.87^{\circ}E$ [13]. Both crashes were essentially attributed to pilots' lack of skill in flying in the mountains, even though those were Air India pilots who regularly flew across the Himalayas and the smaller mountain ranges surrounding the Himalayas. There are claims[14] that Air India 101 carried 41,000 tonnes of fuel when it crashed, the combustion of so much fuel close to the ground

---

[5] http://www.independent.co.uk/news/world/europe/germanwings-plane-crash-andreas-lubitz-supporters-claim-co-pilot-has-been-framed-by-airline-to-cover-10146685.html; https://www.youtube.com/watch?v=7OcR7HZIIRI.

[6] http://sgtreport.com/2015/04/field-mcconnell-pilots-are-not-buying-the-andreas-lubitz-story-it-doesnt-add-up/

[7] http://www.ibtimes.co.uk/germanwings-a320-plane-crash-explosion-smoke-before-airbus-plunged-into-french-alps-1493351

[8] http://en.wikipedia.org/wiki/Air_France_Flight_178

[9] http://en.wikipedia.org/wiki/Crossair_Flight_498; http://www.airliners.net/aviation-forums/general_aviation/read.main/220648/; http://www.breakingnews.ie/world/death-crash-pilot-was-on-tranquillisers-65210.html.

[10] According to http://www.ibtimes.com/germanwings-pilot-andreas-lubitz-was-framed-friends-allege-lufthansa-crash-cover-1864914 antidepressants prescribed to treat a serious "psychosomatic illness" were also found in Andreas Lubitz' home.

[11] http://aviation-safety.net/database/record.php?id=19700221-1, https://en.wikipedia.org/wiki/Swissair_Flight_330

[12] http://en.wikipedia.org/wiki/Air_India_Flight_245 and http://pazhayathu.blogspot.com/2010/06/air-india-most-crashes-could-have-been.html

[13] http://aviation-safety.net/database/record.php?id=19660124-0 and http://en.wikipedia.org/wiki/Air_India_Flight_101

[14] http://www.dnaindia.com/mumbai/report-was-homi-bhabha-s-plane-hit-by-italian-aircraft-1248198

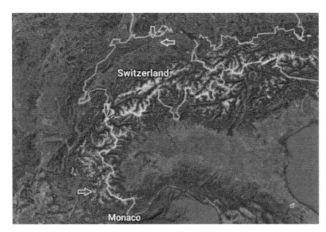

Figure 1: Six airplane crashes in the Alps:
1) Swissair 330 at $47.54^\circ N, 8.24^\circ E$ on 1970/2/21;
2) Crossair 498 at $47.47^\circ N, 8.47^\circ E$ on 2000/1/10;
3) Air India 101 at $45.88^\circ N, 6.87^\circ E$ on 1966/1/24;
4) Air India 245 at $45.83^\circ N, 6.86^\circ E$ on 1950/11/3;
5) Air France 178 at $44.29^\circ N, 6.7^\circ E$ on 1953/9/1;
6) Germanwings 9525 at $44.28^\circ N, 6.44^\circ E$ on 2015/3/24.

Airline and date of crash	Nearest New/Full Moon-closest/2nd closest perigees
Germanwings 9525, 2015/3/24	34 days after 2015/2/18-19 New Moon-2nd closest perigee
Air France 178, 1953/9/1	23 days before 1953/9/23 Full Moon-closest perigee
Crossair 498, 2000/1/10	29 days after 1999/12/22 Full Moon-closest perigee
Swissair 330, 1970/2/21	16 days after 1970/2/5-6 New Moon-closest perigee
Air India 101, 1966/1/24	12 days before 1966/2/5 Full Moon-closest perigee
Air India 245, 1950/11/3	36 days before 1950/12/9 New Moon-closest perigee

Table 1: All six air crashes in the Alps were within 36 days of the nearest New/Full Moon-closest perigee; Germanwings 9525 crash was 34 days after 2015/2/19 New Moon-2nd closest perigee, with the perigee only 115 km farther away than the closest perigee of the year at 356,876 km. The number of days within 36 days of New/Full Moon-closest/2nd closest perigee of the year is $\frac{(2 \times 36 + 1) \times 2}{\text{number of days in a full lunar cycle} \approx 413} \approx \frac{146}{413} < 0.36$; one would expect about $0.36 \times 6 \approx 2$ crashes to be within 36 days of New/Full Moon-closest/2nd closest perigee, not all six. The probability that all six crashes occur within 36 days of New/Full Moon-closest/2nd closest perigee of the year is $0.36^6 < 0.0022 = 0.22\%$. See `https://www.fourmilab.ch/earthview/pacalc.html`.

should have caused a huge fire and a thunder-like explosion; neither was observed suggesting that the aircraft exploded high above the ground at the altitude where the depletion of oxygen would have prevented combustion. There were news reports that at about the same time an Italian aircraft had gone missing, with some speculating that the two planes collided, but no remnants of a second airplane have been found near the crash site of Air India 101. There were, however, two reported crashes of F-104G on 1966/1/24-25 near Rome, and on 1966/1/25 near Accumuli close Rome[15]; the two crashes might have been the same crash reported differently.

Of the six crashes so geographically close to each other: one is attributed to co-pilot's suicide;

---

[15]`http://theaviationist.com/2009/04/21/air-india-101-conspiracy-theory/`; `http://www.916-starfighter.de/F-104_AMI_losses.htm`

4

one to captain's intoxication; one to a bomb from nowhere; two to Air India pilots' inability to fly in the mountains as if they did not get enough experience flying across the Himalayas regularly. The locations of the six airplane crashes form a nice geometric pattern shown in Figure 1. All six crashes occurred within 36 days of New/Full Moon - closest/2nd closest perigee of the year as shown in Table 1, the probability of that is 0.22%.

Nor has the media mentioned that the Germanwings 9525 crash was preceded by two very similar almost-crashes: 2014/12/15 Loganair BE6780[16] was struck by a fireball believed to be ball lightning between Shetland Islands $\approx 60^\circ N, 1^\circ W$ and Aberdeen $\approx 57^\circ N, 2^\circ W$ [17] in UK, autopilot ignored the pilot's commands and sent the plane into the sea at 9,500 feet per minute, merely 1,100 feet above the water the pilot wrestled back control; 2) on 2014/11/5 in Spain Lufthansa 1829[18], while on autopilot, went into a dive reaching 4000 feet/minute speed before the crew was able to regain control of the plane. Autopilot failure, although not that common, is not rare either, e.g. on 2011/7/22 Air France 471 almost crashed in turbulent weather due to autopilot failure (it will be discussed further down in the article); on 2001/7/3 Vladivostok Air 352 crashed at $\approx 52.31^\circ N, 104.3^\circ N$ [19], flight recordings indicated autopilot's emergency shutdown[20]. What is unusual is that within merely five months three aircraft (Lufthansa 1829, Loganair BE6780, Germanwings 9525) suddenly went down in a rather small geographic region encompassing UK, Spain and the Alps. In the wake of the Germanwings crash "a number of crew members refused to fly, though safety concerns were not cited as a factor."[21] What were the crews afraid of? Might the six airplanes have been brought down by ball-lightening-like fireballs, just like Loganair BE6780?

The six crashes occurred in the Alps, where at the end of WWII Allied aircraft pilots observed strange fireballs nicknamed 'foo fighters'[22]. The first 'foo fighters' were observed in November 1944 by a military crew flying from Dijon, France at $47.29^\circ N, 5.04^\circ E$ to patrol the area north

---

[16] http://news.stv.tv/north/303736-plane-hit-by-lightning-while-flying-from-aberdeen-to-shetland-airport/; http://www.dailymail.co.uk/news/article-3034472/Hero-Loganair-pilot-pulls-plane-North-Sea-nosedive-just-SEVEN-SECONDS-spare.html?ITO=1490&ns_mchannel=rss&ns_campaign=1490; http://avherald.com/h?article=4813ed2d.

[17] Exactly 26 years earlier on 1988/12/21 Pan Am 103 exploded over Lockerbie at $\approx 55^\circ N, 3^\circ W$.

[18] http://avherald.com/h?article=47d74074

[19] http://www.irkutsk.org/fed/aircrash2001.html

[20] http://www.tailstrike.com/040701.htm; http://aviation-safety.net/investigation/cvr/transcripts/cvr_vlk352.php

[21] http://www.huffingtonpost.co.uk/2015/04/01/andreas-lubitz-supporters-share-conspiracy-theories-germanwings-plane-crash_n_6983132.html?icid=maing-grid7%7Cuk%7Cdl1%7Csec1_lnk2%26pLid%3D340809

[22] http://en.wikipedia.org/wiki/Foo_fighter; http://sped2work.tripod.com/foo_fighters.html; http://naziufomythos.greyfalcon.us/foofighters.html; War Department Classified Message Center Outgoing Classified Message, 1945/1/2 may be found at http://www.project1947.com/fig/1945a.htm.

of Strasbourg, Germany at $53.5°N, 13.75°E$, while the last ones were seen by pilots in February, 1945 near La Spezia, Italy at $44.1°N, 9.82°E$ [23], the sites of the six crashes are between the two locations. The number of such sightings in late 1940s - early 1950s, not just in Europe but all over the world, was so drastic that to study them the CIA and US air force created the Robertson Panel[24], Projects Sign, Grudge and Blue Book[25]; one of the leading scientists of the latter, Dr J. Allen Hynek, suggested that the fireballs in the sky might be an unknown natural phenomenon[26].

There have been other reports of fireballs or similar objects in the Alps and surrounding areas; although the reliability of many reports is unknown, some seem to be quite reliable[27], and some go back centuries[28]. On 1794/6/16 n Sienna at $43.32°N, 11.33°E$, a rain of iron rocks was observed[29]; while some attributed the rocks to a meteorite, others believed the rocks were lava bombs from Vesuvius, 320 km away at $40.82°N, 14.43°E$ whose eruption started eight hours earlier[30]. On 1857/11/16 a slowly falling orange-size fireball was observed, upon burning out it turned into a small bristling mass of black fibers[31]. On 1808/4/8 over the town of Pignerol, Piedmont, Italy at $44.88°N, 7.33°E$, a loud sound was heard, an earthquake was felt, aerial phenomena and luminous objects were seen in the sky; on 1808/4/19 a stone, currently believed to be a meteorite, fell from the sky near Fidenza, Italy at $44.87°N, 10.07°E$ [32]. On 25-28 July and 7 August of 1566, residents of Basel, Switzerland, at $47.567°N, 7.6°E$ reported seeing an unusual sunrise, a total eclipse of

---

[23] http://naziufomythos.greyfalcon.us/foofighters.html

[24] https://en.wikipedia.org/wiki/Robertson_Panel

[25] https://en.wikipedia.org/wiki/Project_Sign; https://en.wikipedia.org/wiki/Project_Grudge; https://en.wikipedia.org/wiki/Project_Blue_Book

[26] http://www.artgomperz.com/a1998/dec/hynek.htm, part 21d.

[27] Charles Fort, New Lands, part 2, available at 1)http://www.resologist.net/landsei.htm; 2)http://sacred-texts.com/fort/land/index.htm; 3)https://en.wikipedia.org/wiki/Charles_Fort; 5)https://en.wikipedia.org/wiki/New_Lands. Fort's speculations/conclusions cannot be taken seriously; however, many descriptions of fireballs or similar objects seem to be genuine

[28] http://www.theepochtimes.com/n3/703237-nasa-reports-on-credible-ufo-sightings-in-ancient-times/

[29] Surendra Verna, The Mystery of the Tunguska Fireball, ISBN 1 84046 620 0, page 18, https://books.google.com.ph/books?id=c9mhBQAAQBAJ&pg=PA114&lpg=PA114&dq=tunguska+eruption&source=bl&ots=NNqcYwTUDI&sig=fEe4mzHvJoEqLHkw39wOkH_4Ing&hl=en&sa=X&ved=0ahUKEwjvvPjDs-PQAhXEjpQKHcgUD8QQ6AEITzAJ#v=onepage&q=tunguska%20eruption&f=false; http://adsabs.harvard.edu/abs/1995Metic..30R.540M.

[30] Geophysicists usually arrive at the conclusion that a rock is of extraterrestrial origin if it is mostly comprised of iron and nickel. However, the Earth's liquid core is comprised mostly of iron and nickel, and there is nothing preventing the iron and nickel from the Earth's liquid core from travelling through the mantle to a volcano. During an eruption the iron and nickel may be shot into the atmosphere and then land as iron-nickel rocks.

[31] Charles Fort, New Lands, part 2, available at 1)http://www.resologist.net/landsei.htm; 2)http://sacred-texts.com/fort/land/index.htm; 3)https://en.wikipedia.org/wiki/Charles_Fort; 5)https://en.wikipedia.org/wiki/New_Lands. Fort's speculations/conclusions cannot be taken seriously; however, many descriptions of fireballs or similar objects seem to be genuine

[32] Charles Fort, New Lands, part 2, available at 1)http://www.resologist.net/landsei.htm; 2)http://sacred-texts.com/fort/land/index.htm; 3)https://en.wikipedia.org/wiki/Charles_Fort; 5)https://en.wikipedia.org/wiki/New_Lands. Fort's speculations/conclusions cannot be taken seriously; however, many descriptions of fireballs or similar objects seem to be genuine

the moon with a red sun rising, and a cloud of black spheres in front of the sun[33]. On 1561/4/14, residents of Nuremberg, Germany, at $49.45^\circ N, 11.083^\circ E$ reported seeing a large black triangular object and hundreds of spheres, cylinders and other odd-shaped objects that moved erratically overhead. Fireballs have been reported to appear and disappear near and even inside aircraft[34]; since a rock-like fireball can not get inside a solid airplane, at least some fireballs must be moving electric charges creating plasma-like environment around themselves. Almost perfect spheres with high content of iron were recently found in Bosnia[35] near $43.99^\circ N, 18.178^\circ E$; similarly-shaped spheres were found in Otago, New Zealand near $45.345^\circ S, 170.826^\circ E$ and in Hokianga Harbour, New Zealand $35.526^\circ S, 173.379^\circ E$ [36]; also in at $46.967^\circ, 103.45^\circ W$.

If, indeed, fireballs contributed to the crash of Germanwings 9525, what created them? Was there a strong electromagnetic anomaly that could have been responsible for the fireballs? Such anomaly would have affected the flow of cosmic rays. Figure 2 shows periodicity of ≈ 95 days in the maxima of the cosmic ray intensity at the time of the Germanwings 9525 crash. The data on cosmic ray intensity do not go back beyond 1953, but the data on sunspot numbers, correlated to cosmic ray intensity, are available all the way to 1900; the maxima of sunspot numbers show similar periodicity of $\approx 93 - 94$ days in 1943-1946, when 'foo fighters' were observed in Europe. Could the periodicity somehow have resonated with one of the Earth's internal frequencies to produce fireballs[37]? Could the oscillations in the cosmic ray intensity be related to the giant microwave pulse of energy on 2015/3/23-24 described by so many web sites[38]? Such large microwave distortions could be caused by an upper M-class/X-class solar flare or a very large CME from the Sun; but there was neither at the time. The giant microwave pulse was preceded on 2015/3/21 by an unexplained "magnetic short circuit" announced by CERN[39], could it also have been related to

---

[33] https://en.wikipedia.org/wiki/1566_celestial_phenomenon_over_Basel

[34] http://www.nature.com/nature/journal/v224/n5222/abs/224895a0.html; http://www.daviddarling.info/encyclopedia/B/ball_lightning.html. Numerous accounts of such reports may be found in Paul Sagan's book "Ball lightning: paradox of Physics" https://books.google.com.ph/books?id=OLbvX5UnxXoC&pg=PA70&lpg=PA70&dq=fireball+inside++plane&source=bl&ots=eoA50KlxOC&sig=m6eLuY_VntccRNn8cDIdu5jVO20&hl=en&sa=X&ved=0CD8Q6AEwCGoVChMI_Y_iwvHcxwIVx6GUCh03Ggt9#v=onepage&q=fireball%20inside%20%20plane&f=false although the author assumes that all fireball are ball lightening.

[35] https://www.forbes.com/sites/shaenamontanari/2016/04/18/that-massive-stone-sphere-in-bosnia-is-probably-not-from-a-lost-civilization/#21fd1082174e

[36] https://en.wikipedia.org/wiki/Moeraki_Boulders

[37] https://en.wikipedia.org/wiki/Resonance

[38] http://dutchsinse.com/3272015-giant-microwave-pulses-seen-across-europe-africa-and-atlantic-on-march-23-into-24th/; http://tropic.ssec.wisc.edu/real-time/mimic-tpw/global/anim/20150322T000000anim72.gif; http://cultureofawareness.com/2015/03/28/dutchsinse-giant-microwave-pulse-seen-accross-europe-africa-and-atlantic-on-march-23-into-24th/

[39] http://dutchsinse.com/3242015-cern-magnet-short-circuits-today-operations-now-postponed/; http://press.web.cern.ch/; http://www.bbc.com/news/science-environment-32038186.

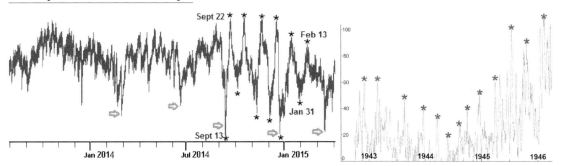

Figure 2: The left frame shows the hourly average of cosmic ray intensity in 2014/9/13 - 2015/2/13; it exhibits periodicity of $\approx 28-29$ days with the minima and maxima marked by black asterisks; and periodicity of ≈ 95 days marked by orange arrows. The right frame shows the daily average of sunspot numbers from late 1943 to early 1946; its maxima, marked with blue asterisks, appear almost periodically every 93-94 days. The 28-29 day periodicity is very unlikely to be due to the lunar motion as the lunar motion is always present yet such a well-pronounced 28-29 day periodicity in cosmic rays is not that common. http://solarscience.msfc.nasa.gov/greenwch/spot_num.txt, http://cosrays.izmiran.ru/.

the oscillations in the cosmic ray intensity?

The conclusion that Lubitz deliberately rammed the plane into the Alps was based on two "key pieces of evidence"[40]:

1) The black box recordings indicate that the captain stepped out of the cabin. There was a loud bang between 9:30 am - 9:34 am, interpreted by investigators as the captain trying to enter the cockpit, accompanied by captain's yells "For God's sake, open the door!"; a loud metallic banging against the cockpit door at 9:35 am, interpreted by investigators as the captain's attempts to break in with a crow bar; the captain shouting: "Open the god damn door" at 9:37 am; noises similar to violent blows on the cockpit door were recorded on five occasions at 9:39 over the course of a minute. Eerily there were no sounds of the captain's voice from 9:33 to 9:37 am; why was the captain silent for the full four minutes? According to some sources[41] "A buzzer requesting access to the cockpit is heard at 09:34. Knocking and muffled voices asking for the door to be opened are heard until the end of the recording." While some sources claim the captain's voice was clearly heard, others claim only "muffled voices". Could the bangs heard on the recordings have been fireballs hitting the airplane? On Airbus jets, access to cockpit may be requested via a keypad outside the cockpit door, in which case a buzzer sounds in the cockpit, and the pilot inside decides whether to unlock the door. If there is no response from the cockpit for a request to open the

---

[40]http://www.dailymail.co.uk/news/article-3016466/Open-goddamn-door-Desperate-final-pleas-Germanwings-captain-emerge-black-box-transcript-reveals-Lubitz-s-repeated-attempts-coax-pilot-toilet.html?ITO=1490&ns_mchannel=rss&ns_campaign=1490;

[41]http://www.bbc.com/news/world-europe-32072218

8

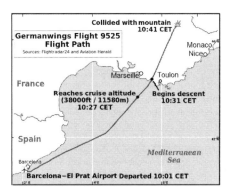

Figure 3: The flight path of Germanwings 9525. The flight started in Barcelona at 09:01 UTC, reached cruising altitude at 9:27 UTC. The plane began its decent at 9:31 UTC just a few miles away from Marseille and Toulon. If the pilot indeed wanted to kill as many people as possible, the logical thing for him would be to direct the aircraft towards Marseille, Toulon, Nice or Monaco. https://commons.wikimedia.org/wiki/File:4U9525_flight_path_v1.svg

door, the flight crew can enter an override code on the keypad, which sets off a 30-second alert in the cockpit[42]. Nowhere in the published reports of the black box recordings could we find any mention of the outside keypad, why? The pilot inside the cockpit may lock the door and render the outside keypad useless but only for five minutes; then the locking of the door must be repeated[43]. Such a process would have left some sounds in the black box recordings, why has this not been mentioned? The second black box has never been found, or so the public has been told.

2) The co-pilot disconnected autopilot at about the time when the aircraft went down. Was it the co-pilot or autopilot of Germanwings 9525 that put the plane into a dive? Lubitz was certainly aware of Loganair BE6780, Lufthansa 1829 and Air France 471, most likely, he was aware of Vladivostok Air 352 as well; switching autopilot off might have been the most logical thing for him to do had the autopilot put the plane into a dive. Might it be that he did not open the cabin door because he fainted, panicked or froze, or simply could not afford to waste time to open it while trying to save the plane?

Numerous articles mention that Lubitz had vision problems possibly psychosomatic[44], yet nowhere could the author of this paper find the description of these problems. Was Lubitz complaining about seeing fireballs in the sky? And if indeed, Lubitz wanted to commit suicide and take as many people with him as possible, why did he ram the plane in a remote area of the Alps rather than Marseille, Toulon, Nice or Monaco, where he could have easily killed hundreds, or even thousands, more? As Figure 3 shows the four cities were within his reach. Some sources[45] claim the aircraft issued a distress call, while others deny it.

On 2015/4/7, just two weeks after the crash of Germanwings 9525, at $\approx 64^\circ N, 22.6^\circ W$, Icelandair 671,[46] was hit by a lightening, leaving a gaping hole in the nose of the aircraft. The crew did not even realize the damage to the plane, as all systems continued to function normally, and proceeded until they safely landed seven hours later. The hit did not show on any instruments to alert the crew, nor was it registered by data recorders. Since such hits were not registered by onboard equipment or felt by the crew, the investigators would not be aware of them either. How many aircraft have been brought down without pilots even realizing what hit them? An interesting analysis of the crash of Germanwings 9525 is provided at `https://www.sott.net/article/294482-Germanwings-crash-Not-the-full-story`, the author also suggests that Germanwings 9525, as well as several others, may have been brought down by fireballs/rocks of extraterrestrial origin.

---

[42] `http://www.latimes.com/la-fg-cockpit-security-20150326-story.html`; `https://qz.com/370386/this-video-shows-how-a-pilot-might-have-been-locked-out-of-the-cockpit-of-germanwings-9525/`.

[43] `http://www.bbc.com/news/blogs-magazine-monitor-32070528`.

[44] `http://www.telegraph.co.uk/news/worldnews/europe/germany/11501771/Andreas-Lubitz-had-eyesight-problem-which-threatened-his-career-as-a-pilot.html`

[45] `https://www.youtube.com/watch?v=UkpptZOgDWw`

[46] `http://www.eturbonews.com/57450/lightning-strikes-icelandair-flight-671`; `http://www.thedenverchannel.com/news/local-news/denver-bound-icelandair-flight-671-struck-by-lightning04082015`; `http://www.denverpost.com/business/ci_27872975/denver-bound-icelandair-flight-from-reykjavik-hit-by`.

2 EgyptAir 990, TWA 800.

> But we thought it was correct at the time. The President
> thought it was correct. Congress thought it was correct.
> Colin Powell, https://en.wikipedia.org/wiki/United_Nations_
> Security_Council_and_the_Iraq_War#cite_note-5

The crash of Germanwings 9525 is almost a mirror image of the crash of EgyptAir 990 at $40.35^{\circ}N, 69.76^{\circ}W$ on 1999/10/31 at $\approx 6:50$ am The cause of the crash was determined to be first officer Al Batouti's suicide. The transcripts from the cockpit voice recorder and the flight data recorder[47] indicate strange sounds, the first 'thunk' sound was at 1:48:22, at 1:48:30 Al Batouti made an unintelligible comment which was followed for the next 48 seconds by the sounds of 'thumps' and 'muffled thumps'[48]. What were the sounds? The captain left the cabin before things went wrong, when he returned, he asked repeatedly: "What's happening, what's happening?" and then said: "What is this? What is this? Did you shut the engines? ... get away in the engines? ... shut the engines". What was the captain referring to by "this"? Al Batouti replied: "It's shut", the final recorded words are the captain's, "Pull with me". There was no indication of a struggle for the control of the aircraft, the recording data show full cooperation of the two pilots. There was no indication of an explosion on board, the engines operated normally for the entire flight until they were shut down. The NTSB concluded that the left engine and some small pieces of wreckage separated from the aircraft before water impact. The autopilot was also disengaged, but was it because the autopilot tried to crash the plane like that of Loganair BE6780? Why would Al Batouti, who was approaching retirement, which he planned to split between a 10-bedroom villa outside of Cairo and a beach house near El Alamein[49], commit suicide? If he planned to commit suicide, why did he buy an automobile tire the day before to bring home to Egypt? Al Batouti's words "I rely on Allah", interpreted by some as a proof of suicide, are not typical of something one say before killing himself, but rather while responding to something unexpected beyond one's control. The crash occurred at the time when CRI exhibited unusual recurrence of

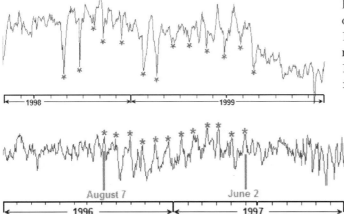

Figure 4: The daily average of cosmic ray intensity in 1998/5/1 - 1999/12/31; the minima, marked by red asterisks, show periodicity of $\approx$ 14.5 days. See `http://cosrays.izmiran.ru/`.

Figure 5: The daily average of cosmic ray intensity for 1996-1997; the maxima, marked by red asterisks, show periodicity of $\approx 27 - 28$ days in 1996/8/7 - 1997/6/2. See `http://cosrays.izmiran.ru/`.

minima/maxima shown in Figure 4.

TWA 800[50] crashed on 1996/7/17 at about 13:45 at $40.66^{\circ}N, 72.63^{\circ}W$, very close to the crash site of EgyptAir 990. Eyewitnesses on the ground and other pilots in the air reported seeing a bright object "streaking" towards the doomed aircraft the US National Transportation Safety Board ruled out an accidental missile strike. Was it a fast moving fireball? The crash occurred at the time when CRI exhibited unusual recurrence of minima/maxima shown in Figure 5 similar to that in Figure 4.

There were several more unexplained crashes in the area, e. g. Fight 441 of US Navy, American Airlines 514; American Airlines 1, American Airlines 1502, New York Airways 972, which we do not discuss here due to space restrictions; their stories, full of unanswered questions, may be found on the Internet.

---

[47] `http://www.thehullthread.com/egyptair.htm`
[48] `http://abcnews.go.com/International/story?id=82910`
[49] `http://www.theatlantic.com/magazine/archive/2001/11/the-crash-of-egyptair-990/302332/`
[50] `http://en.wikipedia.org/wiki/TWA_Flight_800`

3   TransAsia Airways 222, Air Algeirie 5017, China Airlines 611.

> If the man doesn't believe as we do, we say he is a crank, and that settles it. I mean, it does nowadays, because now we can't burn him.
>
> Mark Twain

Here is another collection of events, this time mostly from Asia, suggesting that the fireballs might be related to earthquake lights:

1) TransAsia Airways 222 crashed on 2014/7/23 at 11:06 am at $23.585^\circ N, 119.64^\circ E$ [51] amidst heavy rains. Unusual sounds were recorded by the cockpit voice recorder, interpreted as the sounds of a propeller churning on trees. The crash was accompanied by a magnitude 4.2 earthquake on 2014/7/22 at $\approx 16:47$, at $23.96^\circ N, 122.53^\circ E$ and a magnitude 6.2 earthquake on 2014/8/3 at 08:30 at $27.19^\circ N, 103.41^\circ E$. Merely 3 hours later, on 2014/7/24 at 1:55 Air Algerie 5017 crashed at $14.67^\circ N, 1.95^\circ W$ amidst stormy weather[52] right in the middle of seismic activity shown in Figure 6; the cockpit voice recorder recovered after the crash was unreadable[53], although the data recorder has also been recovered, the author could find no information on what was on it. On 2014/7/23 a person took photo of an ovoid-shaped object in the skies of Rajajipuram area of Lucknow[54] at $26.8^\circ N, 80.9^\circ E$; not too far from Varanasi at $25.28, 82.96^\circ N,^\circ E$, where on on 1798/12/19 a rain of rocks was observed[55]; while currently attributed to a meteorite, the rocks could have been lava bombs from Barren Island which had a series of eruptions in 1789 - 1804. On 2014/7/23, a fireball was observed in Lucknow, India at $26.8^\circ N, 80.9^\circ E$ and a magnitude 4.3 earthquake struck at $26.0^\circ N, 89.7^\circ E$ at 22:21, followed within 74 minutes by two magnitude 4.0-4.2 aftershocks; on 2014/7/22 a a magnitude 4.2 earthquake struck at $23.96^\circ N, 122.53^\circ E$.; on 2014/8/3 a magnitude 6.2 earthquake at $27.19^\circ N, 103.41^\circ E$.

---

[51] http://en.wikipedia.org/wiki/TransAsia_Airways_Flight_222; http://en.wikipedia.org/wiki/China_Airlines_Flight_611

[52] http://en.wikipedia.org/wiki/Air_Alg%C3%A9rie_Flight_5017

[53] http://globalnews.ca/news/1496269/flight-ah5017-air-algerie-cockpit-voice-recorder-unreadable/

[54] http://en.wikipedia.org/wiki/UFO_sightings_in_India#2014; http://indiatoday.intoday.in/story/ufo-aliens-spotted-flying-over-lucknow/1/373893.html

[55] Surendra Verna, The Mystery of the Tunguska Fireball, ISBN 1 84046 620 0, page 21, https://books.google.com.ph/books?id=c9mhBQAAQBAJ&pg=PA114&lpg=PA114&dq=tunguska+eruption&source=bl&ots=NNqcYwTUDI&sig=fEe4mzHvJoEqLHkw39wOkH_4Ing&hl=en&sa=X&ved=0ahUKEwjvvPjDs-PQAhXEjpQKHcgUD8QQ6AEITzAJ#v=onepage&q=tunguska%20eruption&f=false; http://web.mit.edu/redingtn/www/netadv/SP20130218.html.

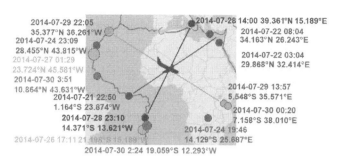

2014-07-29 22:05
35.377°N 36.261°W
2014-07-24 23:09
28.455°N 43.815°W
2014-07-27 01:29
23.724°N 45.581°W
2014-07-30 3:51
10.864°N 43.631°W
2014-07-21 22:50
1.164°S 23.874°W
2014-07-28 23:10
14.371°S 13.621°W
2014-07-26 17:11 21.198°S 15.109°W
2014-07-30 2:24 19.059°S 12.293°W

2014-07-28 14:00 39.361°N 15.189°E
2014-07-22 08:04
34.163°N 26.243°E
2014-07-22 03:04
29.868°N 32.414°E
2014-07-29 13:57
5.548°S 35.571°E
2014-07-30 00:20
7.158°S 38.010°E
2014-07-24 19:46
14.129°S 25.687°E

Figure 6: Earthquakes of magnitude $\geqslant 4$ in July 21-30, 2014. Purple/ pink/blue/khaki/green/brown circles indicate the earthquakes struck within 10.5 hours of each other; for all practical purposes we may assume that the earthquakes of the same color struck simultaneously. The line connecting one earthquake marked purple with the point between the other two, the line connecting earthquakes marked pink, and the line connecting earthquakes marked blue intersect at almost the same point, which is very close to the crash site of Air Algeirie 5017. The line connecting the earthquakes marked green passes close to that point. It is as if one giant earthquake wanted to strike at the point of the intersection but could not make through the thick continental crust and broke up into smaller 'branches' which struck at the colored locations. Brown and khaki circles indicate earthquakes which struck within 10.5 hours of each other, but the lines connecting them do not pass through close to the crash site. The khaki circle indicates a magnitude 6.0 earthquake on 2014/7/7 at 1:29 am at $23.72°N, 45.58°W$. See https://earthquake.usgs.gov/earthquakes/ search/.

2) China Airlines 611 crashed on 2002/3/25 at 7:33 am at $23.99°N, 119.68°E$. It was accompanied by a magnitude 4.3 earthquake on 2002/5/27 at $22.9°N, 121.1°E$; a magnitude 6.1 earthquake on 2002/5/28 at $24.07°N, 122.26°E$; and a magnitude 6.2 earthquake on 2002/5/15 at $24.636°N, 121.922°E$. On 2002/7/22, five gigantic luminous jets between 60-70 km long were observed over the South China Sea from Taiwan[56].

Exactly 18 years prior to the crash of TransAsia Airways 222 and almost six years prior to the crash of China Airlines 611, in the evening of 1976/7/21, several seismologists observed numerous small 'fireballs' coming from the ground[57]. Sixty one minutes later, a magnitude 6.1 earthquake struck on 1976/7/21 at 15:11 at $24.75°N, 98.64°E$. Just a few days later, on the night of J 1976/7/27-28 many people reported strange lights, loud sounds and fireballs flying across the sky around Tangshan, China, $39.6°N, 118.18°E$; on July 27, 1976/7/27 a magnitude 7.5 earthquake struck at $39.6°N, 117.9°E$ [58], followed on 1976/7/28 by a magnitude 7.4 aftershock at $39.66°N, 118.4°E$. On 2010/7/7 in Xiaoshan Airport, Hangzhou, China at $30.25°N, 120.17°E$ a fireball was reported by a flight crew[59]; followed by a magnitude 5.0 earthquake on 2010/7/8

[56]. T. Su, R. R. Hsu, A. B. Chen, Y. C. Wang, W. S. Hsiao, W. C. Lai, L. C. Lee, M. Sato, H. Fukunishi, Gigantic jets between a thundercloud and the ionosphere, Nature, vol. 423, http://sprite.phys.ncku.edu.tw/new/news/ 0626_presss/nature01759_r.pdf

[57] Ol'khovatov, A., On the tectonic interpretation of the 1908 Tungiska event, 2010, may be found at web site http: //olkhov.narod.ru/tunguska.htm; Wallace, R., Teng, T, Prediction of the Sungpan-Pingwu earthquake, August 1976, Bulletin of the Seismological Society of America, 1980 70/4, p. 1199, in Japanese with English abstract.

[58]Chen, Y., Booth, D., The Great Tangshan Earthquake of 1976: An Anatomy of Disaster, 1988, 53, New York: Pergamon Press.

[59]http://abcnews.go.com/International/ufo-china-closes-airport-prompts-investigation/story?id=

14

at $24.42^{\circ}N, 122.11^{\circ}E$ and a magnitude 5.2 earthquake on 2010/7/9 at $24.75^{\circ}N, 122.6^{\circ}E$. The fireballs described in this paragraph may have been earthquakes lights, such lights are capable of traveling/appearing quite far from the earthquake's epicenter.

4 Air France 447, Air France 471, SilkAir 185.

> It is not uncommon for engineers to accept the reality of phenomena that are not yet understood, as it is very common for physicists to disbelieve the reality of phenomena that seem to contradict contemporary beliefs of physics.
>
> H. Bauer

Here is one more collection of three crashes and one almost-crash:

1) Air France 447 crashed on 2009/6/1 at $3.07°N, 30.56°W$ [60]. Right before the crash the flight crew had raised the aircraft's nose, reducing its speed until it entered an aerodynamic stall[61]. Why would the crew do it? The black box recordings indicate the presence of a strange aroma, like an electrical transformer, flooding the cockpit, and a sudden temperature increase[62], as well as a loud sound, interpreted as the sound of slipstream, presumably, due to the accumulation of ice crystals on the exterior of the fuselage. Where did they come from? The behavior of the crew is best described as irrational, why? Some pilots reported seing an "intense flash" in the area where Air France 447 came down[63]. Air France 447 crashed not too far from the city of Colares, Brazil at $\approx 0.94°S, 48.28°W$, known for 'Operação Prato'[64], the sight of numerous fireballs of different shapes and sizes, some luminous and some not; the fireballs were reported to be coming out of water, some exploded in the air producing streams of particles seen as beams of light; there were reports of bluish lights under water. Just 22 hours before the crash, on 2009/5/31 a magnitude 4.9 earthquake struck at $4.548°N, 32.571°W$;

2) On 2011/7/22 close to the crash site of Air France 447, Air France 471 almost crashed in turbulent weather supposedly due to autopilot failure[65], it went into a steep climb and began losing speed; a magnitude 5.9 earthquake struck on 2011/7/27 nearby at $10.8°N, 43.39°W$.

---

[60] http://www.nytimes.com/2011/05/25/world/europe/25france.html; http://en.wikipedia.org/wiki/Air_France_Flight_447; http://aviation-safety.net/database/record.php?id=20090601-0

[61] http://www.bea.aero/fr/enquetes/vol.af.447/point.enquete.af447.27mai2011.en.pdf

[62] http://www.popularmechanics.com/flight/a3115/what-really-happened-aboard-air-france-447-6611877/, if these are indeed the true unedited recordings.

[63] https://www.sott.net/article/186672-What-are-they-hiding-Flight-447-and-Tunguska-Type-Events; http://edition.cnn.com/2009/WORLD/americas/06/04/plane.crash/; http://blog.seattlepi.com/aerospace/2009/06/04/air-france-flight-447-other-pilots-saw-intense-flash-in-sky/

[64] https://en.wikipedia.org/wiki/Opera%C3%A7%C3%A3o_Prato; http://ufos.about.com/od/bestufocasefiles/p/colares.htm; http://www.ufocasebook.com/brazil1977ufoflap.html.

[65] http://www.dailymail.co.uk/news/article-2034685/Air-France-jet-autopilot-fails-drama-echoing-Brazil-crash.html; http://avherald.com/h?article=44280b2a

16

3) AirAsia 8501 crashed on 2014/12/27 at 23:17 close to $3.62^\circ S, 109.71^\circ E$ (almost antipodally to the crash site of Air France 447) amidst bad weather[66]. Before the crash, radar data showed an "unbelievably steep climb"[67] as if the pilot was trying to avoid something ahead, the climb mirrored that of Air France 447. The cockpit voice and data recorders indicated that the captain took "the very unusual initiative to pull the circuit breaker for the FAC, cutting power to it a few minutes before the end of the flight".

4) SilkAir 185 crashed on 1997 /12/19 at $2.46^\circ S, 104.94^\circ E$ [68], exactly 17 years before the crash of AirAsia 8501 and almost at the same place; the crash was ruled to be the captain's suicide. The investigation revealed 'chip-outs' and numerous burrs of unknown origin on the servo valve of the plane's rudder. What could have caused them?

Another famous crash attributed to pilot's suicide is that of LAM Mozambique Airlines 470[69] on 2013/11/29 at $18.19^\circ S, 21.87^\circ E$ amidst heavy rain; the black box indicated sounds as if "some-one pounded on the cockpit door before the crash". It was accompanied by two earthquakes of magnitude $\geqslant 4.5$ on 2013/11/26 at $8.01^\circ S26.736^\circ E$ and on 2013/12/2 at $24.93^\circ S28.611^\circ E$. On 2013/11/29 a fireball was reported close to the crash site[70], exactly 20 years earlier, on 1993/11/18 another fireball was sighted in Sasolburg, South Africa $26.814^\circ S, 27.8286^\circ E$; there were 22 earthquakes of magnitude 3.0 - 4.8 near the site of the sighting; four of them, shown in blue, were of magnitude $\geqslant 4.0$. One more fireball sighted at the end of July, 1997 at $26.48^\circ S, 29.22^\circ E$ accompanied by a magnitude 5.0 earthquake on 1997/7/21 a at $26.86^\circ S, 26.62^\circ E$, followed by two magnitude 4.8 aftershocks on 1997/7/29 at $27.89^\circ S, 26.7^\circ E$ and on 1997/8/1 at $27.94^\circ S, 26.58^\circ E$. The region is home to many unique features: 1) Bushveld Igneous Complex; 2) the submerged freshwater sinkhole Boesmansgat; 3) Vredefort dome; 4) Morokweng crater; 5) graphite deposits in Limpopo; 6) Kimberley, Venetia, Finsch diamond mines; 7) large deposits of rear earth metals and nickel.

[66] http://en.wikipedia.org/wiki/Indonesia_AirAsia_Flight_8501; http://en.wikipedia.org/wiki/SilkAir_Flight_185

[67] http://www.independent.co.uk/news/world/asia/airasia-flight-qz8501-radar-data-shows-unbelievably-steep-climb-before-crash-9951797.html

[68] http://en.wikipedia.org/wiki/Indonesia_AirAsia_Flight_8501; http://en.wikipedia.org/wiki/SilkAir_Flight_185

[69] http://en.wikipedia.org/wiki/LAM_Mozambique_Airlines_Flight_470; http://aviation-safety.net/database/record.php?id=20131129-0

[70] http://www.ufoinfo.com/sightings/southafrica/131129.shtml

5   The Valentich story.

Australian pilot Frederick Valentich and his aircraft vanished on 1978/10/21 at $38.85°S$, $143.52°E$ [71]. Before his disappearance he reported seeing something, asked to identify what it was, Valentich radioed, "It isn't an aircraft" then his transmission was interrupted by unidentified noise described as "metallic, scraping sounds".

On 1991/10/29, exactly 13 years after the Valentich disappearance, a 707-368C of the Royal Australian Air Force unexpectedly stalled and crashed into the sea off East Sale, Victoria at $38.089°S, 147.1494°E$ [72] close to the site of the Valentich disappearance; the crash was attributed to loss of control.

There were four more poorly explained crashes in the region: 1) Australian National Airways Stinson crashed on 1945/1/31 at $37.02°S, 144.5663°E$ [73]; several witnesses reported hearing a sharp crack, when they looked up they saw the airplane spiraling downwards with part of one wing separated from the remainder of the aircraft; 2) on 1946/3/19 an Australian National Airways Douglas DC-3 crashed at $42.86°S, 147.52°E$ [74]; an investigation panel was unable to establish the cause but the mutilated body of a large bird with a wingspan of about 1.8 m found on Seven-Mile Beach a fortnight after the accident lead the investigators to believe that the bird may have struck the aircraft causing the crash; 3) on 1948/9/2 an Australian National Airways DC-3 crashed at $31.511°S, 150.933°E$, not to be confused with the crash described in the previous paragraph. The inquiry found that the probable cause of the crash was interference with the aeroplane's magnetic compass due to a nearby electrical storm and a defect in the navigational signals sent by the Government-maintained Kempsey low-frequency radio range station, an important navigational aid to flights in the area[75]; 4) on November 30, 1961/11/30 Ansett-ANA 325 crashed at $33.9807°S, 151.1998°E$ [76]; the investigation concluded that approximately 9 minutes after take-off the outer section of the right wing had been torn away and the aircraft crashed into Botany Bay amidst the rain, thunder and lightning associated with the thunderstorm over Botany Bay, remarkably no one saw or heard the crash.

---

[71] https://en.wikipedia.org/wiki/Disappearance_of_Frederick_Valentich
[72] http://aviation-safety.net/database/record.php?id=19911029-0
[73] https://en.wikipedia.org/wiki/1945_Australian_National_Airways_Stinson_crash
[74] http://en.wikipedia.org/wiki/1946_Australian_National_Airways_DC-3_crash
[75] https://en.wikipedia.org/wiki/1948_Australian_National_Airways_DC-3_crash
[76] https://en.wikipedia.org/wiki/Ansett-ANA_Flight_325

18

Of Australia's five largest cities, Melbourne, Sydney and Adelaide are close to the location, so the air traffic is heaviest in the region, and a large number of accidents is expected. What is unexpected is the mystery surrounding the accidents described here. The accidents occurred near a group of underwater volcanoes off the southern tip of Tasmania around $44.3^{\circ}S, 147^{\circ}E$.

On 1966/4/6 in Melbourne, Australia, over 200 witnesses saw a UFO described as grey, saucer-shaped, with a slight purple hue, about twice the size of a family car[77]. The location of the sighting at $37.941^{\circ}S, 145.134^{\circ}E$ is close to that of Valentich disappearance, the 1991/10/29 unexpected stall and crash of a Royal Australian Air Force plane, the 1946/3/19 crash of Douglas DC-3.

---

[77] http://en.wikipedia.org/wiki/Westall_UFO

6   Recent events.

> It is really quite amazing by what margins competent but conservative scientists and engineers can miss the mark, when they start with the preconceived idea that what they are investigating is impossible. When this happens, the most well-informed men become blinded by their prejudices and are unable to see what lies directly ahead of them.
>
> Arthur C. Clarke

If, indeed, there is something in the air that brings down airplanes, has it ever been witnessed by other pilots who lived to talk about it? There are a rather large number of such witness reports, here are three coming from near where Loganair BE6780 was struck by a ball lightning, Icelandair 671 was struck by lightening: 1) on 2016/4/17 a British Airways plane was struck by an unidentified flying object on the approach to the Heathrow Airport at $\approx 51.5°N, 0.5°W$, [78]; 2) 2013/7/19 at $\approx$ 18:35, near Heathrow Airport, a pilot reported seeing a silver "rugby ball"-shaped flying object just a few meters away from his jet[79]; 3) on 2012/12/2 near Glasgow, a crew of a passenger plane saw an undetermined flying object that did not show up on radars[80]; not too far from Bala, North Wales at $52.9°N, 3.6°W$ where on 1974/1/23 people witnessed lights accompanying a magnitude 3.5 earthquake, Pembroke at $51.7°N, 5.°W$ where on 1892/8/18 an observation of flying body coincided with a magnitude 5.1 earthquake and Hereford at $52.°N, 2.6°W$ where on 1986/12/17 another observation of a flying body coincided with an earthquake[81]. Here are three reports from other places: 4) on 2016/11/14 a Porter Airlines flight almost collided with an unidentified ball-like flying object about three meters wide near Billy Bishop Toronto City Airport not too far from the crash sites of TWA 800 and EgyptAir 990, to avoid a collision the pilots put the plane in

---

[78]https://www.theguardian.com/technology/2016/apr/28/heathrow-ba-plane-strike-not-a-drone-incident

[70] http://www.huffingtonpost.com/2014/01/06/ufo-jet-airliner-near-miss-over-uk_n_4549399.html. Remarkable is the official description of the events: "The Captain ... *perceived* an object travelling towards them, ... Having very little time to focus,*he was under the apprehension that they were on a collision course* with no time to react. ... The Captain was fully expecting to experience some kind of impact with a conflicting aircraft. ... The Captain *perceived* an object pass within a few feet above the aircraft. ...' The official conclusion was: 'The pilot *was subject to a powerful impression* of immediate danger, *caused by his perception* of an object closing rapidly on his aircraft. ... After some discussion it was decided that, ... the overall dearth of information relating to the event rendered a meaningful finding impossible. It was not possible to trace the object or determine the likely cause of the sighting."

[80]http://www.ibtimes.co.uk/ufo-plane-glasgow-scotland-463308#

[81]http://adsbit.harvard.edu//full/2006A%26G....47e..11M/E000011.000.html

20

a controlled dive[82]; 5) on 2013/6/4 near Chengdu an Air China plane was hit in the air by an unknown ball-like object[83]; 6) on 2014/3/19 near Perth and almost antipodally to the crash sites of EgyptAir 990 and TWA 800, the crew of an airplane sighted a bright strobe light directly in front of the aircraft coming from cylindrically-shaped object towards the aircraft; the pilot took evasive action to avoid a collision with the object[84].

[82]https://www.thestar.com/news/canada/2016/11/14/porter-plane-in-near-miss-with-drone.html; http://www.therecord.com/news-story/6965695-2-hurt-as-porter-flight-evades-mid-air-object/
[83]http://www.dailymail.co.uk/news/article-2339139/Was-bird-A-Plane-Or-UFO--Chinese-passenger-jet-hits-mysterious-object-26-000ft-lands-severely-dented-nose-cone.html; http://cayodagyo.blogspot.com/2013/06/some-possibilities-of-object-that-had.html
[84]http://www.atsb.gov.au/media/4897226/AO-2014-052%20Final.pdf

7 Discussion.

> ... when you have eliminated the impossible, whatever
>
> remains, however improbable, must be the truth ...
>
> Conan A. Doyle,   The Sign of the Four,   chapter 6.

It is hard to tell what exactly caused the airplane crashes described here. The results of the investigations of the crashes seem to be very subjective. Could some of the crashes described here have been misdiagnosed? Very possibly. The crashes of Germanwings 9525, EgyptAir 990, TransAsia Airways 222, LAM Mozambique Airlines 470, Air France 447, and the Valentich flight were preceded by strange sounds in the cabin, interpreted by investigators based on very little information available. The remnants of SilkAir 185 revealed the presence of 'chip-outs' and numerous burrs on the servo valve of the plane's rudder, the cause of which has never been determined; the flaperon discovered on Reunion Island in July 2015 and attributed to the lost Malaysia Airline MH370 had unexplained jagged edges[85]. What caused them? Could the crashes have been caused by electromagnetic fields of/impacts with fireballs, ball lightening, sprites[86], blue jets[87], earthquakes lights[88], volcano lightnings, lava bombs[89], or eddy currents generated by relatively rapidly changing magnetic fields[90] by disrupting the autopilots and onboard computers? Our suggestion is not the first one; the presence of some objects in the sky was advocated by Charles Fort in early 1900s[91], the possibility that the fireballs seen in the sky were a natural phenomenon was proposed by aforementioned Dr. J. Allen Hynek as well as the Robertson Panel, Projects Sign, Grudge and Blue Book[92]. The solar activity is known to affect the Earth's magnetic field while the cosmic ray intensity is known to affect and be affected by the Earth's magnetic field; could

---

[85]http://www.theweek.co.uk/mh370/57641/mh370-plummeted-out-of-sky-at-up-to-20000ft-a-minute

[86]https://en.wikipedia.org/wiki/Sprite_(lightning)

[87]ttps://en.wikipedia.org/wiki/Upper-atmospheric_lightning#Blue_jets

[88]For example, Ouellet, M., Earthquake Lights and Seismicity, Nature, 1990, pp. 348-492.

[89]http://en.wikipedia.org/wiki/Volcanic_bomb;   http://www.geology.sdsu.edu/how_volcanoes_work/ Thumblinks/Lavaball_page.html; Anderson, R., Gathman, S., Hughes, J., Sveinbjörn, S., Bjornsson, S., Jónasson, S., Blanchard, D., Moore, C., Survilas, H., Vonnegut, B., Electricity in Volcanic Clouds: Investigations show that lightning can result from charge-separation processes in a volcanic crater, Science, 1965, 148/3674, pp. 1179-1189, abstract at http://www.sciencemag.org/content/148/3674/1179.abstract?ijkey= 375fba217260b8d42e6453dea57a8c13a4c52918&keytype2=tf_ipsecsha

[90]https://en.wikipedia.org/wiki/Eddy_current

[91]http://www.bahaistudies.net/asma/book_of_the_damned.pdf; http://www.resologist.net/landsei.htm

[92]https://en.wikipedia.org/wiki/Project_Sign; https://en.wikipedia.org/wiki/Project_Grudge; https://en. wikipedia.org/wiki/Project_Blue_Book.

22

then the intensity of the solar and cosmic rays activity somehow affect the appearance, frequency and power of the fireballs? Earthquakes are known to be accompanied by changes in ionosphere[93], could these changes produce fireballs and/or lead to airplane crashes in some other way?

Did the pilots, accused of crashing their planes, actually do it? Maybe, or maybe not. The crash of Germanwings 9525 was attributed to suicide within two days of the crash,[94] further "investigation" marching towards only to confirm the conclusion; the media frenzy and numerous "experts" wholeheartedly supported the conclusion[95]. Dead pilots cannot argue. We are aware of the danger to air travel by volcanic ash[96] and flocks of geese[97], but could the flocks of fireballs/flying rocks/electromagnetic disturbances pose a similar or even greater threat? If there is even a tiny grain of truth in our reasoning, then more planes will be shot down from the sky and more people will die, unless the causes are properly identified and dealt with.

[93] http://www.tau.ac.il/~colin/research/EarthQukes/Workshop/Natan%202010.pdf
[94] For comparison, it took about a year for a Ford dealership to determine why the ABS indicator on the author's Ford Escape was lighting up.
[95] http://www.telegraph.co.uk/news/worldnews/germanwings-plane-crash/11513967/Second-black-box-confirms-French-Alps-crash-co-pilot-Andreas-Lubitz-acted-deliberately.html
[96] https://en.wikipedia.org/wiki/British_Airways_Flight_9; https://en.wikipedia.org/wiki/KLM_Flight_867; http://pubs.usgs.gov/fs/fs030-97/.
[97] https://en.wikipedia.org/wiki/US_Airways_Flight_1549

Printed in Great Britain
by Amazon